Wild & Furry Animals
of the
NORTHERN APPALACHIAN MOUNTAINS
From Pennsylvania to Maine

Written and Illustrated
by
Lee James Pantas

R. Brent and Company
Johnson City, Tennessee

Wild & Furry Animals of the Northern Appalachian Mountains.

Published in Johnson City, Tennessee
by R. Brent and Company
1301 Lake Drive Ext
Johnson City TN 37601
828-231-3923
robbin@rbrent.com

Editor and publisher: *Robbin Brent*
Cover & interior design: *Jane Ware*
Design editor & compositor: *Rick Soldin*
Illustrations: *Lee James Pantas*

Publication Data

Pantas, Lee James
wild & furry animals of the northern appalachian mountains
p. cm.
ISBN Paperback: 978-0-9910398-7-6

Lee James Pantas may be contacted at 828-779-1569 or leepantas@bellsouth.net

IMAGE CREDITS

- Cover illustration: *Canada Lynx*; Name: Lee James Pantas: leepantas.com
- Image: *Clownfish*, an extract from the last painting in *A Journey to Liza's Reef*: "The Island of Vanuasemia," p. 7; Name: Lee James Pantas: leepantas.com
- Back-cover illustrations: *Moose* and *American Badger*; Name: Lee James Pantas: leepantas.com

This book is dedicated to my son, Daniel,
and to my daughter, Susanna.

Contents

Introduction

The ancient mountains of the Northern Appalachians, a vast and wonderful place, are home to many species of extraordinary and interesting wild mammals. Although I now live in the Southern Appalachian Mountains near Asheville, North Carolina, parts of my youth were spent in Vermont and New York. During those formative years I spent countless hours roaming the forests and meadows of this mountainous region that stretches from Pennsylvania to the northernmost tip of Maine and on into Canada.

This book is a sister volume to my *Wild & Furry Animals of the Southern Appalachian Mountains*. And, like that book, this volume documents by illustration and brief descriptive text all of the more common, as well as some of the rarer and lesser-known, creatures found in the region. It is inspired in part by memorable encounters with many of the creatures as a youth, not in the least a magical unforgettable moment when my path crossed that of a magnificent Canada Lynx on an isolated snowy road in northern Vermont. It is also inspired by numerous early childhood memories of watching wide-eyed and graceful White-tailed Deer racing across snow-covered fields on my grandparent's farm near Canisteo in upper New York State.

At a time when wild habitats are vanishing at an alarming rate and many species of animals are in danger of extinction, my hope is that this book will educate and inspire others, especially young people, to appreciate, protect, and respect the wild animals of our world.

An act of kindness toward an animal is a step closer to God.

About the Author

Lee James Pantas, born in 1941 and originally from Connecticut, has lived just outside Asheville, North Carolina, since 1989. He has a master's degree in Ecology (University of Vermont, 1966), and served as a Captain in the United States Army Military Police from 1967 to 1969. In 1978 he founded New Covenant House, a Catholic soup kitchen in Stamford, Connecticut, to feed the homeless. He also coached hurdles and sprints for the A.C. Reynolds Middle and High Schools Track & Field teams in Asheville, North Carolina, from 1993 to 2023.

In addition to the volume you are holding in your hands, Lee is the author and illustrator for three other books: *The Ultimate Guide to Asheville & the Western North Carolina Mountains, Wild & Furry Animals of the Southern Appalachian Mountains*, and *A Journey to Liza's Reef*. The *Clownfish* image below was extracted from *Liza's Reef* painting: "The Island of Vanuasemia."

Early in his art career, Lee completed more than 2,500 architectural pen & ink drawings of private residences, historic buildings, inns, and churches. Currently he focuses on drawings of wild animals, as well as painting. He is well known for his imaginative paintings of interstellar vistas, coral reefs, and exotic endangered animals and their habitats. His work is found in many private and public collections.

Correspondence should be directed to:

Lee James Pantas
18 Garren Mountain Lane
Fairview NC 28730

leepantas@bellsouth.net
828-779-1569
LeePantas.com

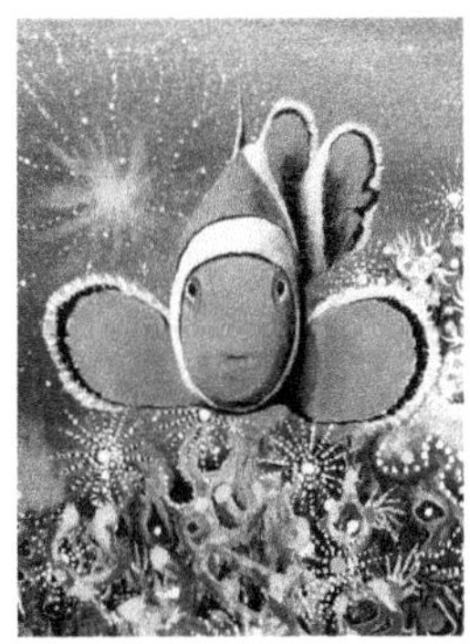

Wild & Furry Animals of the Northern Appalachian Mountains

The wild and furry animals presented in this book are found in an immense and beautiful region of America referred to as the Northern Appalachian Mountains. This magnificent territory encompasses the Alleghany Mountains in Pennsylvania, the Blue Ridge and Bear Pond Mountains in Maryland, the Ramapo Mountains of New Jersey, the Catskill and Adirondack Mountains of New York, the Taconic Mountains of Connecticut, the Green Mountains of Vermont, the Berkshire Hills Mountains of Massachusetts, the White Mountains of New Hampshire, and the Longfellow Mountains of Maine.

This book presents all of the mammals, both large and small, that are likely to be encountered in these mountains, as well as the some of the rarest and least likely to be seen.

American Badger

American Badger

Taxidea taxus

What is their preferred habitat?

Open grasslands, fields, and pastures.

How big are they?

Males weigh between 14 and 19 pounds, and females up to 16 pounds.

What do they like to eat?

Primarily carnivorous, they prey on mice, voles, woodrats and, occasionally, on snakes, lizards, amphibians, birds, skunks, and insects.

What is their average lifespan in the wild?

Usually 9 to 10 years and, in some cases, up to 14 years.

What is their fur color?

Their body fur is grizzled brown, black, and white, and their triangular faces have a distinctive white stripe flanked by brown or blackish "badges" on the cheeks.

Fun Facts

- American badgers have stocky and low-slung bodies with short powerful legs and huge foreclaws.
- They are ferocious predators and are members of the *Mustelidae*, a diverse family of carnivorous mammals that include wolverines, otters, ferrets, and weasels.
- American badgers occasionally pair up and work with coyotes as a team while hunting, usually one badger to one coyote, but occasionally up to three coyotes with one badger.
- They do not hibernate but may become less active in winter. They live in a den network of tunnels and underground chambers.
- They are primarily nocturnal hunters but occasionally hunt during the daytime.

American Black Bear

American Black Bear

Ursus americanus

What is their preferred habitat?

Uninhabited forests.

How big are they?

Large males weigh between 100 and 500 pounds but can grow to 600 pounds. The smaller female typically does not exceed 200 pounds.

What do they like to eat?

They are omnivorous and eat a wide variety of foods, including acorns, roots, insects, plants, berries, fruit, fish, and small mammals. They also will kill and eat young deer, elk, or sheep.

What is their average lifespan in the wild?

Usually 10 to 20 years.

What is their fur color?

Most commonly black, but also can be brown and dark brown.

Fun Facts

- American black bears are found only in North America and are the smallest of the three bear species found on the continent.
- They have short, non-retractable claws that give them excellent tree-climbing abilities.
- They are highly dexterous and capable of opening door latches and screw-top jars.
- Depending on local weather conditions and food supply, most hibernate in the winter.
- American black bears can easily outrun humans, with tops speeds of 25 to 30 miles per hour. They can run uphill and downhill with equal speed.
- They are good swimmers and will readily enter water in search of fish.
- Their sense of smell is much superior to that of dogs.
- Some black bears in British Columbia are pure white in color and are referred to as "Spirit Bears" by the local Native Americans.

American Ermine

American Ermine (Short-tailed Weasel)

Mustela erminea richardsonii

What is their preferred habitat?

Various habitats include forests, wetlands, grasslands, and mixed woodlands.

How big are they?

Between 5 to 13 inches long. They usually weigh less than a pound, and males are twice as large as females.

What do they like to eat?

They are opportunistic hunters and prefer to prey on other small mammals, but will also feed on insects, worms, and berries.

What is their average lifespan in the wild?

Normally 4 to 6 years.

What is their fur color?

The upper parts of their bodies are brown, while the lower parts are white to yellowish white, except in winter when their coats are white. The tip of the tail is black.

Fun Facts

- They are also called *stoats*.
- In the winter months, from December through April, their coats turn all white.
- American ermines frequently use the nests of its prey for shelter, adding dried plant materials, fur, and even feathers to the nest lining. They often have multiple nests.
- Few predators are able to kill them because of their agility and aggressiveness.
- Occasionally they are prey to owls, hawks, coyotes, and long-tailed weasels.
- They are largely solitary creatures and get together only for the breeding season in July and August. Males will occasionally bring freshly killed prey to the female of a family group as a way of gaining access to the group.

American Marten

American Marten

Martes americana

What is their preferred habitat?

Northern forests with mature coniferous and other trees.

How big are they?

American martens are usually 14 to 17 inches long with a tail that measures from 7 to 9 inches long. They usually weigh between 2 and 4 pounds, and the males are larger than the females.

What do they like to eat?

Their favorite prey species are red squirrels. They also eat voles, mice, snowshoe hares, and even birds, eggs, insects, and berries.

What is their average lifespan in the wild?

Normally 12 to 17 years.

What is their fur color?

They have gray heads, dark brown or black tails, light brown backs, and cream-colored patches on their chest.

Fun Facts

- American martens are adept swimmers and excellent climbers.
- They look cute and cuddly but are fierce predators. They usually kill their prey, like many other larger predators, with a bite to the back of the neck.
- American martens are solitary, nocturnal creatures, and are active year-round.
- They are very curious animals and actively investigate every crevice, stump, or fallen log in search of food.
- Their most vicious enemies are fishers, bobcats, lynx, hawks, and owls.

American Mink

American Mink

Neogale vison

What is their preferred habitat?

Forested areas near rivers, streams, marshes, swamps, ponds, and lakes.

How big are they?

Males are typically 13 to 18 inches long and weigh from 1 to 3 pounds, the females slightly smaller.

What do they like to eat?

Favorite prey are fish but they also feed on crustaceans, amphibians, rodents, and birds.

What is their average lifespan in the wild?

Normally 1 to 3 years.

What is their fur color?

Dark brown to black.

Fun Facts

- American minks are graceful swimmers that can dive as deep as 16 feet underwater. They swim by using their back legs.
- They run on land in a distinctive bounding gait, and can climb trees with ease.
- American minks are solitary creatures.
- They are usually able to catch fish after a furious 5-to-20-second chase.
- James Audubon, the famous naturalist, once reported seeing a mink carrying a 12-inch trout.
- They kill snakes but do not eat them.

Big Brown Bat

Big Brown Bat

Eptesicus fuscus

What is their preferred habitat?

They are found in every habitat, although most abundant in deciduous forest areas.

How big are they?

The large ones have wingspans up to 13 inches.

What do they like to eat?

All types of insects. Their preferred food is beetles.

What is their average lifespan in the wild?

Up to 6 years.

What is their fur color?

Reddish brown with black nose, ears, wings, and feet.

Fun Facts

- The big brown bat is very common in North America.
- They are nocturnal and hunt at night, and often use echolocation to locate their prey.
- They are significant predators of agricultural pests. This directly benefits farmers and many use bat boxes to attract them to their property.
- One of their favorite foods is the cucumber beetle.
- Although rabies is common in all bats, research has shown that it is rarer in big brown bats, which are immune to the disease.
- During the day they roost in sheltered areas, such as caves, tunnels, tree cavities, and man-made structures.
- They hibernate in winter, often within 50 miles of their summer roosts. They tolerate cold weather fairly well.

Bobcat

Bobcat

Lynx rufus

What is their preferred habitat?

Woodlands and open-canopy forests.

How big are they?

Bobcats are about twice as big as an ordinary housecat, and generally weight between 11 and 28 pounds, with females smaller than males. There have been rare reports of bobcats reaching 60 pounds.

What do they like to eat?

Rabbits and hares are number one on their list, but they also eat birds, mice, and many other small animals.

What is their average lifespan in the wild?

Normally 7 years.

What is their fur color?

Beige to brown, with spotted or lined, dark brown or black, markings.

Fun Facts

- Bobcats are named for their short, bobbed tails.
- Bobcats are nocturnal, solitary animals and are rarely seen by humans. Their growls are deep and fearsome, sounding as if they were coming from a much larger animal.
- They are fierce stealth hunters and can kill animals much bigger than themselves, including deer.
- They can leap over 10 feet and are excellent swimmers.
- They are the most abundant wild cat in America and can have home ranges as large as 20 square miles.

Canada Lynx

Canada Lynx

Lynx canadensis

What is their preferred habitat?

Cold, northern, moist coniferous forests with snowy winters. They prefer to be near rocky areas, bogs, and swamps.

How big are they?

Adults range from 19 to 22 inches high at the shoulder, and usually weigh between 11 and 37 pounds.

What do they like to eat?

Their main prey animal is the snowshoe hare, which makes up to 90% of their diet. They also eat birds, meadow voles, carrion, and occasionally deer.

What is their average lifespan in the wild?

Between 10 to 20 years.

What is their fur color?

Grayish-brown mixed with pale-brown fur on the back, and buff or grayish-white on the stomach, legs, and feet.

Fun Facts

- Canada lynx are nocturnal hunters and can spot prey up to 250 feet away in the dark. They stalk their prey before pouncing on it. They also wait in "ambush beds" on trails for their prey to come by.
- They are excellent swimmers and tree climbers and often wait on branches for passing prey.
- They have long legs and unusually large paws that work like snowshoes, allowing them to easily hunt and travel in deep snow.
- Canada lynx have distinctive long black tufts on triangular ears, and a black tip on their short tails.
- Since they rely heavily on the snowshoe hare as their main food source, their population rises and falls depending on the local hare population.

Coyote

Coyote

Canis latrans

What is their preferred habitat?

Woodlands.

How big are they?

Between 20 and 50 pounds.

What do they like to eat?

They are omnivores and will eat almost anything. They actively hunt and eat rabbits, fish, frogs, rodents, and other small mammals, as well as insects, snakes, and fruit.

What is their average lifespan in the wild?

Normally 4 to 6 years.

What is their fur color?

Light gray to reddish brown, with areas of black and white.

Fun Facts

- Coyotes are one of the most adaptable and cunning canines on the planet and can be found in a wide variety of habitats, from deserts to alpine meadows, and even in large cities like New York and Los Angeles.
- They are social animals and prefer to live and hunt in packs.
- Coyote howls, referred to by scientists as group yip-howls, are short howls that rise and fall in pitch, punctuated with staccato yips, yaps, and barks. These group yip-howls are often mistaken for a large pack of animals all raising their voices at once. Because of the variety of sounds produced by each coyote, and the way sound is distorted as it passes through the environment, two coyotes can sound like six to eight animals.
- When walking or running, the coyote carries its tail downwards rather than horizontally as wolves do.
- They are very strong swimmers.
- They can and do breed with wolves and dogs.
- Coyotes raise their pups in dens or holes in the ground.

Deer Mouse

Deer Mouse

Peromyscus maniculatus

What is their preferred habitat?
Woodlands and meadows with plenty of ground cover.

How big are they?
Small, their bodies are only 3 to 4 inches long.

What do they like to eat?
They are omnivorous and eat primarily seeds, flowers, nuts, fruits, leaves, fungi, and insects, including spiders and caterpillars.

What is their average lifespan in the wild?
Around 1 year.

What is their fur color?
Grayish brown with white underbellies.

Fun Facts

- Deer mice are accomplished climbers, jumpers, and runners, and their name "deer mice" refers to their athletic abilities as well as their fur coloration, which is like that of deer.
- They are nocturnal creatures and spend their days hidden in the nests they build in burrows, hollow trees, or logs.
- Female deer mice construct the nests from grasses, roots, wool, thistle down, moss, and a variety of other natural materials.

Eastern Chipmunk

Eastern Chipmunk

Tamias striatus

What is their preferred habitat?

Forests and woodlands with plenty of ground cover.

How big are they?

Small, their bodies are only 2 to 6 inches long.

What do they like to eat?

They are omnivores with a diet consisting primarily of seeds, fungi, nuts, fruit, and grass.

What is their average lifespan in the wild?

Normally 2 to 3 years.

What is their fur color?

Reddish-brown on their sides, and white on the stomach, with black and white striping on sides and back.

Fun Facts

- Eastern chipmunks have cheek pouches that allow them to carry large amounts of food items to their burrows for storage and consumption. One eastern chipmunk was observed gathering more than 160 acorns in one day. In just two days they can collect enough food to last for the winter.
- They have been known to store up to 8 pounds of food in their burrows.
- Eastern chipmunks dig extensive burrows underneath or next to natural or manmade cover, including logs, stumps, and rocks.
- They are curious animals and can be coaxed into eating out of a human hand.
- A group of chipmunks is called a *scurry*.
- North America is home to 21 species of chipmunks.
- Eastern chipmunks hibernate in the winter but don't sleep all the way through the season. They retreat to their burrows in late fall but wake every few days, raise their body temperatures to normal, feed on stored food, and then go back to sleep.

Eastern Cottontail Rabbit

Eastern Cottontail Rabbit

Sylvilagus floridanus

What is their preferred habitat?

Meadows, open grassy areas, forest clearings, and old fields with abundant grass and plants for food and shrubs for cover.

How big are they?

Between 1 and 3 pounds.

What do they like to eat?

They are herbivores and eat a variety of grasses, clover, leaves, flowers, bark, and seeds.

What is their average lifespan in the wild?

Normally 1 to 2 years.

What is their fur color?

Speckled brown-gray and reddish brown.

Fun Facts

- Eastern cottontail rabbits do not dig burrows, but instead nest in shallow, slanted holes that they dig under clumps of grass or shrubs, and then line with grass and fur.
- They are very active animals, but also can remain sitting completely still for 15 minutes at a time.
- Eastern cottontail rabbits thrive as a species because they reproduce up to five times per season. A single pair could potentially produce 350,000 offspring in just five years if there were no predators.
- They can leap up to 15 feet when necessary to escape predators. When pursued, they circle their territory and jump sideways to break their scent trails.
- Their name is derived from their white, fluffy tail that resembles a ball of cotton.
- On midwinter nights, eastern cottontail rabbits have been observed frolicking on crusted snow. These playful gatherings are believed to be reactions after periods of forced inactivity.

Eastern Gray Squirrel

Eastern Gray Squirrel

Sciurus carolinensis

What is their preferred habitat?

Forests and woodlands, preferably those with oak, walnut, hickory, and other nut-bearing trees.

How big are they?

Between 1 to 2 pounds.

What do they like to eat?

Acorns, nuts, seeds, tree buds, bark, and any food found in residential bird feeders.

What is their average lifespan in the wild?

Normally 8 to 9 years.

What is their fur color?

Gray.

Fun Facts

- Eastern gray squirrels are easily distinguished from the larger fox squirrels by their predominantly gray color. Eastern gray squirrels also have both black- and white-colored variations. Genetic variations also include black squirrels with white tails, and gray squirrels with black tails.
- They are scatter-hoarders, animals that store food in numerous caches for later recovery during winter months. Individual eastern gray squirrels can make up to several thousand caches each season. They often lose up to 25% of their buried food to fellow squirrels or birds, and they don't always find all of their buried nuts, which results in the growth of more trees.
- To prevent other animals from finding their food caches, they sometimes use deceptive behavior, like pretending to bury a food item when they think they are being watched.
- Eastern gray squirrels can find food buried beneath a foot of snow. Once they find the scent of food, they dig a tunnel through the snow to the buried treasure.
- Most eastern gray squirrel populations in major city parks were introduced by humans.
- They are one of the few mammals that can descend a tree headfirst.
- When threatened, they run away in a zigzag pattern. This is an extremely useful strategy in escaping predators, but unfortunately does not work well when they are on roads trying to avoid being run over by cars.
- They construct their dens as nests from twigs and leaves in tree branches, or within the hollow trunks of trees.
- Their front teeth never stop growing and they must gnaw to keep their teeth at the right length.

Eastern Mole

Eastern Mole

Scalopus aquaticus

What is their preferred habitat?

Eastern moles prefer moist, loose soils found in fields, meadows, pastures, and wooded areas.

How big are they?

Typical size is around 6 inches in length and they usually weigh only a couple of ounces.

What do they like to eat?

Earthworms but they also will eat many other types of food, including snails, centipedes, insects, grubs, and slugs.

What is their average lifespan in the wild?

Normally 6 years.

What is their fur color?

Dark to light shades of silver gray.

Fun Facts

- Eastern moles spend 99% of their time underground.
- They have tiny, degenerative eyes with fused eyelids that are limited to distinguishing only between shades of light and dark.
- They have enormous hairless front feet with curved claws that they use for digging.
- They dig burrows just under the surface for foraging. When digging these burrows, they push excess soil up onto the surface referred to by scientists as "molehills."
- Eastern moles can dig tunnels at a rate of more than 15 feet per hour.
- One of their favorite habitats are golf courses.
- Each day eastern moles must eat 25 to 100% of their own weight in food to survive. One mole can eat over 50 pounds of worms in a year.
- They can smell in stereo. This ability helps them to not only detect odors, but also to determine the direction from which the odors are coming.

Eastern Spotted Skunk

Eastern Spotted Skunk

Spilogale putorius

What is their preferred habitat?

Forests and wooded areas.

How big are they?

They are smaller than Striped skunks and usually weigh between 1 and 2 pounds.

What do they like to eat?

They are omnivores and prefer rabbits, small rodents, insects, birds, fruit, and eggs.

What is their average lifespan in the wild?

Normally 1 to 2 years.

What is their fur color?

Black and white.

Fun Facts

- Eastern spotted skunks are rare in the Northern Appalachian Mountains, and only a very small endangered population exists in southeastern Pennsylvania.
- Eastern spotted skunks have 4 to 6 stripes on their backs, giving them a spotted appearance. The only true spot they have is one found on their foreheads. Their distinctive black and white appearance is a warning to potential predators.
- They are excellent climbers and can scurry up and down trees like squirrels.
- Before spraying a potential predator, they stand repeatedly on their front paws in a handstand as a warning display. If that doesn't work, they then assume a horseshoe shaped stance before spraying their foul-smelling musk toward the predator. They have an accurate range of over 10 feet.
- They are nocturnal animals and are usually only seen by humans at dawn or dusk.

Elk

Elk

Cervus canadensis

What is their preferred habitat?
Forests and meadows.

How big are they?
Elk cows average between 500 to 600 pounds, while larger bulls average from 600 to 1,000 pounds. Mature bulls stand 5 feet high at the shoulder, and are up to 8 feet in length.

What do they like to eat?
Grasses, acorns, plants, leaves, and bark.

What is their average lifespan in the wild?
Normally 10 to 20 years.

What is their fur color?
In summer, reddish brown, and light tan throughout the rest of the year.

Fun Facts

- Elk in the Northern Appalachian Mountains are found only in Elk, Cameron, Clinton, Potter, and Clearfield counties of Pennsylvania.
- The largest herd of elk, over 1,400 strong, is found in Elk County. One of the best ways to see these magnificent creatures is to visit the Elk Country Visitor Center in Benezette PA.
- Elk, also known as *Wapiti*, are one of the largest species in the world-wide deer family.
- Only male elk have antlers, which are shed each winter. Antlers may reach 4 feet in length and in summer can grow more than an inch a day.
- Another species of elk—Sika Deer (*Cervus nippon*)—lives in the quiet marshes and forested wetlands on the eastern shore of Maryland. They are small (70-90 pounds) elk that look like deer with white spots on their back and rump. They are native to Japan and were introduced in 1916.
- Bulls have a distinctive vocalization known as *bugling*, which can be heard for miles. These calls are used mainly during mating season and females are attracted to the males that bugle most often and loudest.

Fisher

Fisher

Marts pennanti

What is their preferred habitat?
Fishers prefer coniferous forests, but also are found in mixed hardwood forests.

How big are they?
Typical size is between 29 and 47 inches in length, and they usually weigh between 3 and 15 pounds.

What do they like to eat?
Fishers are omnivores and their prey includes snowshoe hares, eastern cottontail rabbits, porcupines, and other small animals. They also eat insects, fruit, nuts, and mushrooms.

What is their average lifespan in the wild?
Fishers can live up to 10 years.

What is their fur color?
Medium to dark brown, with black legs and a long tapering tail. They also have gold and silver on their head and shoulders, and occasionally a cream patch on their chests.

Fun Facts

- Fishers are one of the few mammals that prey on porcupines.
- Male fishers are twice as big as females.
- They are active day and night and although sometimes known as Fisher Cats, they are not felines nor do they fish.
- Fishers are one of the largest members of the weasel family.
- Fishers use hollow trees, stumps, holes in the ground, and brush piles as "resting sites" during all times of the year. During the winter, fishers also use snow dens that are burrows under the snow.

Fox Squirrel

Fox Squirrel

Sciurus niger

What is their preferred habitat?

Forests and woodlands, especially those with oak, hickory, walnut, and pine trees.

How big are they?

Usually between 17 and 27 inches long, with an 8-to13-inch tail. They typically weigh around 2 pounds.

What do they like to eat?

Favorite foods are acorns, other types of nuts, seeds, tree buds, bulbs, roots, and insects.

What is their average lifespan in the wild?

Normally 3 to 4 years.

What is their fur color?

Gray to tawny brown, with reddish-brown to pale gray underneath.

Fun Facts

- Fox squirrels are the largest species of tree squirrels native to North America.
- They have two types of shelters that they use: tree dens and leaf nests. They prefer to use tree dens during the winter and for raising young because dens provide better insulation.
- Great climbers and jumpers, they can leap over 15 feet in horizontal jumps from one branch to another.
- Fox squirrels can find food buried beneath a foot of snow. Once they find a scent to buried food, they dig a tunnel through the snow to the cache.
- Like gray squirrels, fox squirrels bury nuts in the ground to feed on during the winter.
- Fox squirrels are scatter-hoarders, animals that store food in numerous caches for later recovery, especially during winter months. Individual squirrels can make up to several thousand caches each season.
- They may lose up to 25% of their buried food to fellow squirrels or birds. They also do not find all of their buried nuts, which results in the growth of more trees.
- They are one of the few mammals that can descend a tree headfirst.

Gray Fox

Gray Fox

Urocyon cinereoargenteus

What is their preferred habitat?

Habitats with lots of woods and brush.

How big are they?

Larger than the red fox, gray foxes typically weigh between 7 and 15 pounds.

What do they like to eat?

They are omnivorous and their favorite prey is the eastern cottontail rabbit. They also eat voles, shrews, birds, crickets, grasshoppers, and fruit.

What is their average lifespan in the wild?

Normally 6 to 8 years.

What is their fur color?

Peppery-gray on their backs, and reddish-brown on chest, sides and legs.

Fun Facts

- The gray fox's range includes both North and South America.
- They have the ability to climb trees, unusual for a canid, a skill shared only with the Asian raccoon dog. They have strong claws that allow them to climb to escape predators, including coyotes and domestic dogs, or to access arboreal food sources. They also can jump easily from branch to branch.
- Gray foxes sometimes take naps in trees.
- They make dens in hollow trees, stumps, or burrows to hide from predators and to raise their young.
- Gray foxes are easily differentiated from the red fox by their lack of "black stockings" on their lower legs, and have predominately gray fur, not orange.
- They have vertical pupils and can retract their claws, two physical characteristics of cats.
- Female foxes are called *vixens*.
- Gray foxes have whiskers on their face and legs, which helps them navigate.
- They have superior hearing and have been reported to be able to hear a watch ticking over 40 yards away.
- They are nocturnal and hunt primarily at night.

Groundhog

Groundhog (Woodchuck)

Marmota monax

What is their preferred habitat?

Open meadows and the edges of woodlands.

How big are they?

Usually between 4 to 13 pounds, but have been recorded to weigh over 30 pounds.

What do they like to eat?

Wild grasses and other vegetation. Alfalfa, dandelion, clover, and coltsfoot are some of their favorite grasses.

What is their average lifespan in the wild?

Normally 3 to 6 years.

What is their fur color?

Grayish brown.

Fun Facts

- Groundhogs are excellent swimmers.
- Related to squirrels, they also can climb trees.
- They have two coats of hair that gives them a distinctive "frosted" appearance.
- Groundhogs dig extensive burrows, sometimes more than 60 feet long. These are used for hibernating in the winter, sleeping, and nurseries, as well as a safe retreat when danger is present. There are usually 2 or more entrances.
- February 2 in the United States and Canada is celebrated as Groundhog Day. According to tradition, if it is sunny when it emerges from its burrow, the groundhog will supposedly see its shadow and retreat back into its den, and winter weather will persist for 6 more weeks. If it is cloudy and it sees no shadow, then the spring season will arrive early.
- One of their most common nicknames, "woodchuck," has nothing to do with wood. It is thought to be a corruption of the Native American words *wejack*, *woodshaw*, or *woodchoock*.
- The largest Groundhog Day celebration is held in Punxsultawney, Pennsylvania, the home of its famous resident groundhog, Punxsultawney Phil.

Least Weasel

Least Weasel

Mustela nivalis

What is their preferred habitat?

Least weasels are adaptable and thrive in a multitude of habitats from forests to meadows.

How big are they?

Least weasels are the smallest of the weasel family and usually are only 8 to 9 inches long, and weigh only a few ounces.

What do they like to eat?

Primarily small rodents, and occasionally rabbits, birds, fish, and frogs.

What is their average lifespan in the wild?

Normally 1 to 2 years.

What is their fur color?

For most of the year, light brown with a white underbelly. In winter, their fur changes to all white.

Fun Facts

- Least weasels are found in Maryland, Pennsylvania, and New York State in the Northern Appalachian Mountains.
- The least weasel is the smallest carnivorous predator in the world. Despite its small size, the least weasel is a fierce hunter and capable of killing a rabbit ten times its size.
- Because of its size and very active lifestyle, the least weasel must eat 40 to 60% of its body weight every day.
- In killing its prey, their method is like that of a jaguar. After jumping on its prey, the least weasel wraps its muscular body around the animal to immobilize it, and then delivers a single fatal bite to the back of the head, puncturing the skull or spinal cord.
- They will hunt and kill anything that moves and look like prey, even if they are fully fed. They often kill much more than they can eat. Any carcasses not consumed are stored in underground caches near their dens to eat at a later time.
- The coat of the least weasel turns white in the winter to blend in with the snow. This is characteristic of all weasels. This white winter fur glows a bright lavender color under ultraviolet light.
- Least weasels are solitary animals and make nests in crevices and roots of trees, and abandoned burrows.

Little Brown Bat

Little Brown Bat

Myotis lucifugus

What is their preferred habitat?

Near swamps, ponds, lakes, rivers, and streams.

How big are they?

Very small, usually between 2.5 and 4 inches long, and weighing no more than an ounce.

What do they like to eat?

Insects, especially aquatic insects, gnats, midges, flies, moths, wasps, and beetles.

What is their average lifespan in the wild?

Normally 6 to 7 years.

What is their fur color?

Uniformly brown.

Fun Facts

- The little brown bat is one of the most common bats in North America.
- Evening forages as a group are done usually above bodies of water.
- Little brown bats sleep and groom during the day and hunt at night, usually only for a few hours.
- They use echolocation to find their prey.
- They have three different roosting sites: day roosts and night roosts during spring, summer, and fall; and hibernation roosts in the winter. Day roosts are typically found in trees or buildings, wood piles, and occasionally in caves. Night roosts tend to be in the same area as day roosts but are in tighter, closed-in spaces so they can huddle together for warmth.

Long-tailed Weasel

Long-tailed Weasel

Neogale frenata

What is their preferred habitat?

They prefer to live near a water source and are found in a wide variety of habitats, including farmlands, open areas, woodlands, and thickets.

How big are they?

Adults range from 8 to 10.5 inches in length.

What do they like to eat?

Mice, voles, chipmunks, rabbits, and occasionally birds, reptiles, amphibians, and insects.

What is their average lifespan in the wild?

They can live up to 5 years.

What is their fur color?

They are brown on their backs and sides, and yellowish white on their bellies. Their tails have a black tip. In the northern parts of their range, their fur turns white in winter.

Fun Facts

- Long-tailed weasels have to eat up to 40% of their body weight every day. They are ferocious hunters and often attack other animals larger than themselves.
- They burrow in the ground, live in hollow logs, rock piles, or under barns, but they may also take over the abandoned burrows of other animals.
- Their eyes are black in daylight but glow bright emerald-green at night in the light of a flashlight.
- In the winter when its coat is white, the long-tailed weasel has a unique defense against its enemies: coyotes, foxes, bobcats, and lynx. It uses its black-tipped tail as a distraction and causes the predator to attack its tail, allowing it to escape.

Meadow Vole

Meadow Vole

Microtus pennsylvanicus

What is their preferred habitat?

Moist grassland areas with thick protective cover.

How big are they?

Usually between 5 to 8 inches long, including the tail.

What do they like to eat?

Grasses, sedges, and other vegetation, including many agricultural plant species.

What is their average lifespan in the wild?

Less than 1 year.

What is their fur color?

Dark brown above and silvery-gray below.

Fun Facts

- Meadow voles are one of the most common and prolific small mammals in North America.
- They can swim but have never been seen climbing.
- Depending on the type of habitat, the local population of meadow voles varies widely, from only a few individuals an acre to more than 1,480 per acre.
- Hawks sitting on telephone and electrical wires overlooking pastures are often watching for meadow voles below.
- They are prolific breeders and can have up to 17 litters a year, with the average litter size between 4 to 8 babies.

Moose

Moose

Aleces alces

What is their preferred habitat?
Northern forests. Mixture of mature and young trees, which provide abundant forage.

How big are they?
Six feet at the shoulder and weighing more than 1,000 pounds.

What do they like to eat?
Twigs, leaves, and other woody material from trees. During the summer they also eat large quantities of aquatic plants.

What is their average lifespan in the wild?
Moose can live more than 20 years.

What is their fur color?
Dark brown. Females have a light brown face, and a patch of white fur beneath the tail.

Fun Facts

- Moose are the largest members of the deer family.
- The name *Moose* is an Algonquin Indian term that means *eater of twigs*.
- Both males and females have flaps of skin called *dewlaps* hanging from their throats, and males grow massive antlers up to six feet across during the spring and summer months.
- Moose are highly susceptible to climate change. Warmer winters cause higher tick infestations that can cause moose to die from blood loss and anemia, which results in a decline in their population.
- Females, called *cows*, are highly aggressive in defending their young. They have been known to kill bears, wolves, and even people in defense of their babies.

Muskrat

Muskrat

Ondatra zibethicus

What is their preferred habitat?

Wetlands and swamps, and along streams, rivers, canals, lakes, and ponds.

How big are they?

Usually between 2 and 5 pounds, and 1 to 2 feet long.

What do they like to eat?

Cattails, water lilies, and other vegetation. In winter they will also each insects, crayfish, fish, and freshwater mussels.

What is their average lifespan in the wild?

Normally 1 to 3 years.

What is their fur color?

Medium to dark brown or almost black.

Fun Facts

- Muskrats are not related to rats, although their appearance is similar. They are named for their musky scent, which is emitted from glands at the base of their tails.
- Muskrats are semi-aquatic and spend much of their time in water. They are great swimmers and can swim underwater, backwards and forwards, for over 15 minutes. Their main means of propulsion in the water are their flattened, black scaly tails, which function as rudders.
- For shelter, muskrats build burrows into banks with underwater entrances along streams and lakes. If they live in marshes and swamps, they build 3-to-5-feet-high lodges made of vegetation and mud, also with underwater entrances.
- The inner chamber of burrows and lodges always rise above the surface of the water in order to provide dry shelter for the muskrats.
- Muskrats do not hibernate and are active year-round, feeding at all times of the day.
- Muskrats gather food and carry it to feeding platforms that they build for a place to eat. These platforms are flat, elevated piles of mud and vegetation constructed outside of their burrows. If abandoned, these platforms also are used as nesting sites by many birds.
- Most of the time muskrats mate underwater, and their social life is built around the colony, which typically consists of an adult pair and 2 to 6 young.

North American Beaver

North American Beaver

Castor canadensis

What is their preferred habitat?
Swamps, marshes, streams, lakes, and ponds.

How big are they?
They typically weigh between 20 and 70 pounds. Very old individuals have been recorded at over 100 pounds.

What do they like to eat?
Leaves, buds, and inner bark of live trees, especially aspens, poplar, beech, birch, willow, and alder.

What is their average lifespan in the wild?
Normally 10 to 20 years.

What is their fur color?
Chestnut brown to almost black.

Fun Facts

- The North American beaver is the official animal of Canada.
- They, like all beavers, are monogamous.
- This beaver is the largest rodent in North America and is the 2nd largest in the world.
- Their large front teeth, which they use constantly for gnawing trees and branches, never stop growing. These teeth are orange due to the iron contained in the enamel, a substance that also makes them incredibly strong and sharp.
- During the Ice Ages, North American beavers, known as *Castoroides*, grew up to 8 feet long and weighed over 200 pounds.
- They are best known for constructing dams across streams to create ponds. The dams are constructed from mud, rocks, vegetation, and branches from trees the beavers cut down. They also construct lodges equipped with underground entrances in the middle of the ponds created by their dams.
- Their dams can be enormous structures, with one dam—under construction since the 1970s in northern Alberta—measuring over 2,500 feet long.
- They were brought to the brink of extinction in 1838 North America because of their fur, which was highly sought after for clothing and hats. Much of the westward expansion in America was driven by the quest for beaver fur. With protection in the late 19th and early 20th centuries, their population in America has rebounded to around 15 million.

North American River Otter

North American River Otter

Lontra canadensis

What is their preferred habitat?
Lakes, rivers, streams, swamps, and marshes.

How big are they?
Between 11 and 30 pounds.

What do they like to eat?
They are carnivores, with a diet that includes fish, crayfish, amphibians, clams, mussels, snails, and turtles.

What is their average lifespan in the wild?
Normally 8 to 9 years.

What is their fur color?
Light brown to black.

Fun Facts

- North American river otters swim by propelling themselves with their powerful tails and flexing their long bodies and webbed feet. They can hold their breath underwater for up to 8 minutes.
- They are renowned for their sense of play and love to slide down snow-covered or muddy banks into the water. They are equally at home on land or in the water.
- North American river otters are extremely social animals and form large social groupings, typically from 4 to 8 individuals, with some groups as large as 20.
- They live in dens constructed in the burrows of other animals, or in naturally occurring hollows under logs or river banks.
- They have specialized vision that enables them to view potential meals while swimming in dark or murky water. They easily spot food not visible to other animals. They possess an exceptional sense of smell that also helps them locate prey.
- Unlike most marine mammals, North American river otters do not have a layer of insulating blubber. Instead, air is trapped in their dense fur, which helps to keep them warm.

Northern Flying Squirrel

Northern Flying Squirrel

Glaucomys sabrinus

What is their preferred habitat?

Coniferous and mixed coniferous forests.

How big are they?

About 10 inches from their nose to the tip of their tail.

What do they like to eat?

Primarily acorns, nuts, fungi, and lichens. They also will eat tree sap, insects, bird eggs, and nestlings.

What is their average lifespan in the wild?

Normally 4 to 6 years.

What is their fur color?

Light brown or cinnamon, with creamy white fur underneath.

Fun Facts

- Flying squirrels do not actually fly, but instead glide by opening a membrane called a *patagium* to catch the air. They have been observed to glide up to 75 feet from tree to tree, with their average glide around 15 feet.
- They are strictly nocturnal and generally nest in the holes of trees, preferring dead trees of larger diameter.
- It is believed that they use triangulation to estimate the distance of the landing area when they launch themselves into the air.
- They are social animals and have been observed flying and foraging together in large groups.
- Size is one way to distinguish between the southern and northern flying squirrels. The southern flying squirrels average 8 to 10 inches, while the northern variety are usually 10 to 12 inches long.
- Humans imitate flying squirrels using special suits called *wingsuits*, which are designed after the squirrel's anatomy with its extended air-catching flaps of skin. These wingsuits allow the experienced wearer, jumping from high mountainous elevations and off cliffs, to glide just like squirrels over long distances.

Northern Short-tailed Shrew

Northern Short-tailed Shrew

Blarina brevicauda

What is their preferred habitat?
Grasslands, marshy area, gardens, and forests, especially those with moist leaf litter or thick plant cover.

How big are they?
Between 4 to 5 inches long, and weighing around an ounce.

What do they like to eat?
Primarily carnivorous, they eat insects, earthworms, voles, mice, snails, salamanders, and other shrews.

What is their average lifespan in the wild?
Normally 1 to 2 years.

What is their fur color?
Silvery slate gray, occasionally black or brownish black.

Fun Facts

- The northern short-tailed shrew is notable in that it is one of the few venomous mammals in the world. Their saliva contains a chemical that paralyses their prey and is toxic enough to kill small animals several times larger than the shrew. Their saliva is not strong enough to kill a human, but will cause a very painful bite.
- They are voracious eaters and must consume up to three times their body weight each day.
- Their eyes are degenerated and vision is thought to be limited to the detection of light.
- They hunt using echolocation and a developed sense of touch as they tunnel along below ground, through snow, or leaf litter.
- In the winter, northern short-tailed shrew are active in the airspace between the ground and snow cover.
- Nests lined with fur and leaves are built in underground chambers or under logs or rocks.
- A small pile of old snail shells on the ground might be a tell-tail sign of a northern short-tailed shrew nest.

Norway Rat

Norway Rat (Brown Rat)

Rattus norvegicus

Black Rat

Rattus rattus

What is their preferred habitat?

Both Norway rats *(shown in illustration)* and Black rats prefer to live near human habitation.

How big are they?

Norway rats are usually around 10 inches long. Black rats are slightly smaller and can grow to 7 inches.

What do they like to eat?

Both species are omnivores and will eat almost anything. Cereal grains and seeds are a substantial part of their diet.

What is their average lifespan in the wild?

Both species live for 1 year or less.

What is their fur color?

Norway rats have brown to dark gray fur. The fur of the black rat is primarily black, but also can be tawny brown in color.

Fun Facts

- Norway and Black rats are some of the best known and most common rats in the world. By population and territory occupied, they qualify as the second most successful mammal on the planet after humans.
- Both species live wherever humans live, especially in urban areas. It is estimated there is one rat for every person in the United States. Black rats were once the dominant species in America, but now rank 2nd behind brown rats.
- Both species sleep together and commonly groom each other. They dig extensive burrows with multiple levels of tunnels when soil conditions permit that provide shelter, nest areas, and food storage sites. They have a very powerful chain of command and the largest and strongest rats get the best food and shelter places.
- A group of rats is called a *mischief of rats*.

Porcupine

Porcupine

Erethizon dorsatum

What is their preferred habitat?

Coniferous, deciduous, and mixed forests.

How big are they?

Usually 2 to 3 feet in length, and up to 20 pounds.

What do they like to eat?

They are herbivores and eats leaves, twigs, bark, and green plants.

What is their average lifespan in the wild?

They can live up to 18 years.

What is their fur color?

Brownish-yellow to black, with white highlights on their quills.

Fun Facts

- They have approximately 30,000 hollow quills on their backs and sides that are used for defense. Quills are specially adapted hairs with barbed tips on the ends.
- When threatened, they draw up the skin on their back so that the quills bristle and face in all directions. They keep their backs to the predator and moves its tail back and forth.
- Contrary to popular belief, porcupines cannot throw quills at attackers. They only attack when threatened.
- If a porcupine hits an animal with its quills, they become embedded in the animal.
- Porcupines are mostly nocturnal and do not hibernate.
- Although they can climb trees, porcupines are not especially good climbers. One study found that 30% of one population studied had healed fractures from falling from trees.

Raccoon

Raccoon

Procyon lotor

What is their preferred habitat?

Forests and woodlands.

How big are they?

Between 8 and 23 pounds.

What do they like to eat?

They are omnivores with an opportunistic diet and will eat a wide variety of plant and animal foods. Favorite foods are bird eggs, fish and shellfish, frogs, birds, fruit, insects, nuts, and seeds.

What is their average lifespan in the wild?

Normally 2 to 3 years.

What is their fur color?

Various shades of gray, with a distinctive brownish-black-and-white facial mask and tail.

Fun Facts

- Raccoons are noted for their intelligence, and researchers have found that they are able to remember solutions to tasks for up to 3 years.
- Raccoons are considered smarter than cats and just below monkeys on the mammal intelligence scale.
- They make their dens in hollow trees, as well as abandoned burrows.
- They communicate with each other using over 50 different sounds and up to 15 different calls. They can purr, whistle, growl, scream, and hiss.
- They are known for rinsing their food in water prior to eating it. When no water is available, raccoons will still rub the food to remove debris. Their name, *Procyon lotor*, means *washer dog*.
- Raccoons are not related to dogs, but are close relatives to bears.
- The name *raccoon* is derived from the Algonquin Indian word *arakun*, which means, "he who scratches with hands."
- Raccoons are nocturnal animals, and one theory about their black mask around their eyes is that it helps the animals with their night vision.
- They have 5 fingers on their dexterous front paws, much like human hands, and are very skillful in opening shells, trash cans, doors, and collecting food. Their hands are so nimble they can unlace a shoe, unlatch cages, and even retrieve coins and small objects from pockets.
- They have the ability to climb down trees head first by rotating their hind feet 180 degrees.
- Raccoons are savage fighters and, if cornered, can kill an attacking dog.

Red Fox

Red Fox

Vulpes vulpes

What is their preferred habitat?

Forests and grasslands.

How big are they?

Usually between 6 and 24 pounds.

What do they like to eat?

They are omnivores, but primarily feed on small rodents, birds, insects, and other smaller mammals. They readily eat plant material that includes berries, apples, grapes, grasses, and nuts.

What is their average lifespan in the wild?

Normally 2 to 4 years.

What is their fur color?

Orange-red fur on their heads, backs, and sides, white under the neck and chest, and a long bushy tail with a white tip. They have black ears, legs, and feet.

Fun Facts

- Red foxes are great jumpers and can leap over 6-foot high fences.
- They are denning animals and dig burrows in which they live and raise families.
- Their bushy tails aid them in balance.
- They have vertical pupils and can retract their claws, two physical characteristics shared with cats.
- They stalk their prey like cats, getting as close as they can and then pouncing. They have excellent hearing, especially for the low-frequency sounds produced by rodents digging underground.
- Female foxes are called *vixens*.
- Red foxes have whiskers on their face and legs, which help them navigate.

Red Squirrel

Red Squirrel

Tamiasciurus hudsonicus

What is their preferred habitat?
Coniferous forests.

How big are they?
Adults are usually between 11 and 14 inches from their nose to the tip of their tail.

What do they like to eat?
They feed primarily on the seeds of conifer cones. They can adjust their diet to eat berries, bird eggs, and other types of nuts.

What is their average lifespan in the wild?
They can live up to 8 years.

What is their fur color?
Reddish-brown with a white belly. They have bright white rings around their eyes.

Fun Facts

- Red squirrels are best known for their aggressive personality.
- Known also as the pine squirrel, chickaree, or boomer.
- They have a sweet tooth and are known to occasionally bite into maple trees so they can eat the sap.
- They are easily to identify. They are much smaller than both gray and fox squirrels, but larger than chipmunks.
- They have bushy dark red tails that help them with balance in the trees.
- Red squirrels collect and hide seeds and nuts as food storage for the winter months.

Snowshoe Hare

Snowshoe Hare

Lepus americanus

What is their preferred habitat?
Northern coniferous forests with a dense brushy undergrowth.

How big are they?
They are larger than cottontail rabbits typically weighing between 3 to 4 pounds.

What do they like to eat?
Grass, ferns, leaves, flowers, twigs, and bark. They occasionally will eat carrion.

What is their average lifespan in the wild?
Up to 1 year or less.

What is their fur color?
White during winter months, and rusty to grayish brown during the rest of the year.

Fun Facts

- Snowshoe hares are solitary nocturnal animals and feed only from dusk to dawn.
- The major predator of snowshoe hares is the Canada lynx.
- Their seasonal adaptation of fur turning white in the winter helps them to evade predators by blending in with the snow. Even though most of their fur turns white, the tips of the ears always remain black.
- Snowshoe hares got their name because of the large size of their hind feet. These prevent them from sinking into the snow when they walk and hop.
- Snowshoe hares can jump almost 10 feet horizontally, if necessary, to escape predators.
- They also can run at speeds up to 25 miles per hour in short bursts. They are good swimmers and will also jump into ponds and streams to escape predation.

Southern Bog Lemming

Southern Bog Lemming

Synaptomys cooperi

Northern Bog Lemming

Synaptomys borealis

What is their preferred habitat?

Both the southern bog lemming *(shown in illustration)* and the northern bog lemming prefer bogs, swamps, and marshy areas in forests where the soil is moist and loose and covered with a thick mat of shrubby vegetation.

How big are they?

Both are small, usually weigh less than an ounce, and average 4 to 5 inches in length.

What do they like to eat?

Leaves, stems, seed heads, roots of sedges and grasses, and berries.

What is their average lifespan in the wild?

Normally 1 to 2 years.

What is their fur color?

Reddish to dark brown with pale gray underbelly.

Fun Facts

- The southern bog lemming is quite abundant throughout the Appalachian Mountains, while the northern species is rare and found only in northern New Hampshire and Maine.
- Both species are very similar in appearance and difficult to distinguish. The northern species has rusty-tipped fur at the base of the ears, and the females have eight mammae as opposed to six for the southern species.
- Bog lemmings are small voles that dig tunnels and runways for nesting, feeding, resting, and food storage.
- They are different from other voles in that they like to make small piles of grass next to their runways.
- Contrary to popular lore, bog lemmings are not suicidal and do not throw themselves off of cliffs.
- Both species are legendary for their population fluctuations, living in colonies that can number from a few up to 30.
- They excrete bright green fecal pellets due to the large amount of green vegetation they consume.

Southern Flying Squirrel

Southern Flying Squirrel

Glaucomys volans

What is their preferred habitat?
Deciduous forests that have abundant nut trees.

How big are they?
Smaller than the northern flying squirrel, around 8 to 10 inches.

What do they like to eat?
They are omnivores and prefer nuts and acorns. They will also eat seeds, fungi, fruit, insects, eggs, birds, and carrion.

What is their average lifespan in the wild?
Normally 4 to 5 years.

What is their fur color?
Gray brown on top, and a cream color underneath.

Fun Facts

- Southern flying squirrels are nocturnal animals and are rarely seen by humans.
- They do not fly but instead glide using a membrane called a *patagium* that extends to either side of their body that helps to capture air. They have been observed gliding more than 150 feet.
- It is believed that they use triangulation to estimate the distance of the landing area when they launch themselves into the air.
- They nest in natural tree cavities and woodpecker holes, or build nests out of twigs and leaves.
- They are social animals and have been observed flying and foraging together in large groups.
- Size is one way to tell the southern and northern flying squirrels apart. The southern flying squirrels average 8 to 10 inches, while the northern varieties are usually 10 to 12 inches long.
- Humans imitate flying squirrels using special suits called *wingsuits*, designed after the squirrel's anatomy with their extended flaps of skin. These wingsuits allow the experienced wearer, jumping from high mountainous elevations and off cliffs, to glide just like the squirrels over long distances.

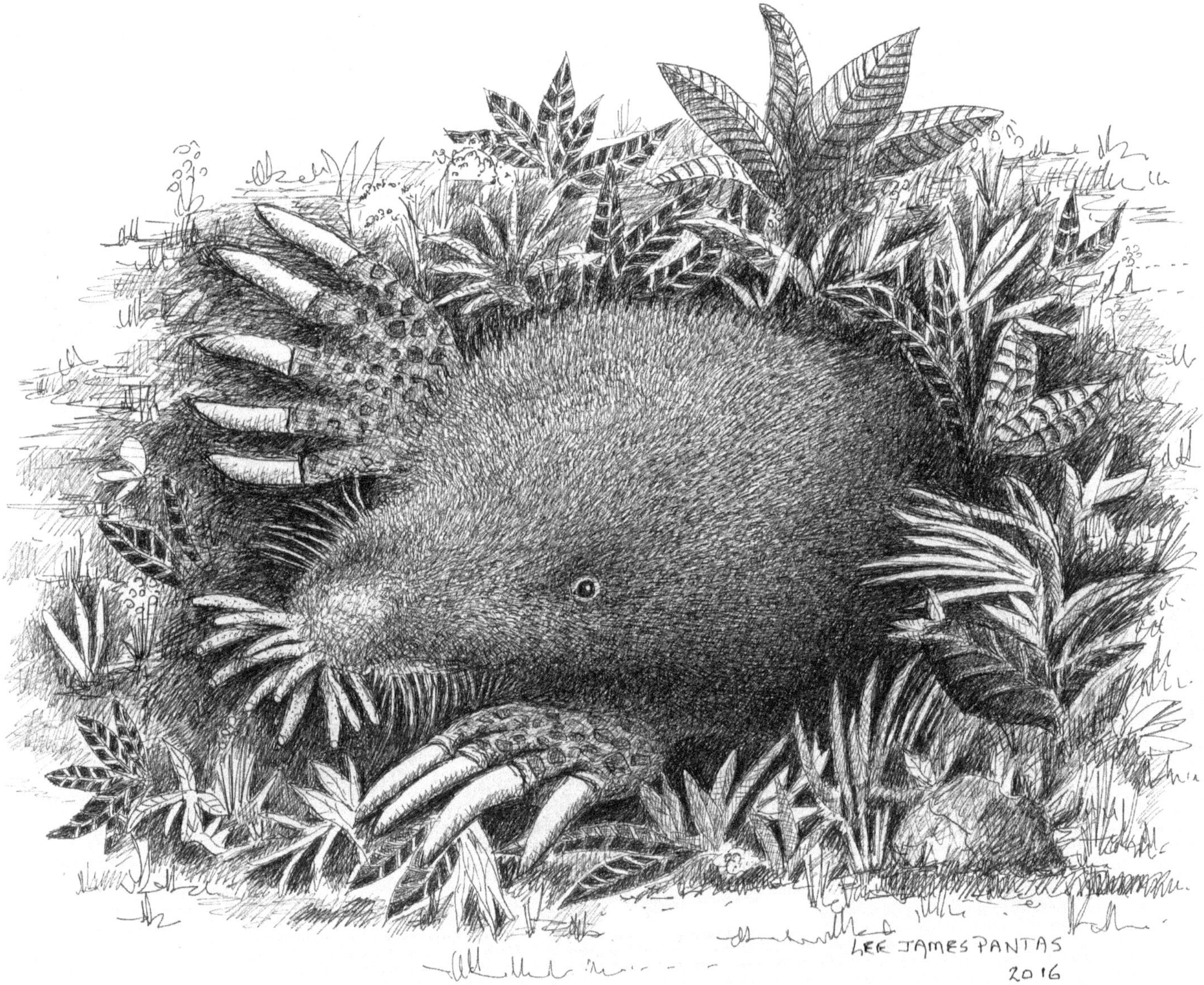

Star-nosed Mole

Star-nosed Mole

Condylura cristata

What is their preferred habitat?
Wet, poorly drained areas, marshes, and the banks of streams, lakes, and ponds.

How big are they?
Size ranges between 5 to 8 inches long, and weighing about 2 ounces.

What do they like to eat?
Invertebrates, aquatic insects, worms, amphibians, mollusks, and fish.

What is their average lifespan in the wild?
Not known for sure, but probably less than 2 years.

What is their fur color?
Blackish brown.

Fun Facts

- The star-nosed mole is easily identified by the 22 pink fleshy appendages or rays that ring its snout and serve as a touch organ. There are more than 25,000 minute sensory receptors, known as Elmer's organs, in these rays, which help the mole feel its way as it digs underground.
- Most people would agree they are one of the world's strangest looking animals.
- They dig shallow tunnels for foraging that often end up underwater.
- They are excellent swimmers.
- They have been recognized by *The Guinness Book of World Records* as the world's fastest forager. This ability is due to their amazingly sensitive star nose that instantly tells the mole whether or not it has found food. It takes them as little as 230 milliseconds to identify and eat prey.
- They have a unique ability to smell underwater. They do this by exhaling air bubbles onto objects they encounter and then inhaling the scented bubbles back again.

Striped Skunk

Striped Skunk

Mephitis mephitis

What is their preferred habitat?
Woodlands, and brushy fields and meadows.

How big are they?
Usually between 2 to 14 pounds.

What do they like to eat?
Primarily an insectivore, they will also eat vegetation, mice, voles, and the eggs and chicks of ground-nesting birds.

What is their average lifespan in the wild?
Normally 2 to 3 years.

What is their fur color?
Black with a white stripe extending back from the head, which divides the shoulders. Typically, they also have a white stripe on their foreheads.

Fun Facts

- The striped skunk is one of the world's most recognizable animals and is often featured in children's books and cartoons. Their black and white coloration is a warning to potential predators not to get too close.
- Like all skunks, they possess scent glands under their tail. The glands contain an oily, yellow, foul-smelling musk that is sprayed when threatened by predators.
- Striped skunks are easily tamed, kept as pets, and also have been extensively bred for their fur.
- They sometimes dig their own dens, but will readily use those abandoned by other animals. They also live in old tree stumps, under logs and rocks, and even in attics and basements of houses.
- In the 19th century, striped skunks were often kept in barns by farmers to kill mice and rats.
- They are nocturnal animals and hunt for food only at night.
- They are very intelligent animals, and one of their more interesting behaviors is their practice of rolling a furry caterpillar in the dirt to get rid of its irritating hairs before eating it. They also can easily get into garbage cans, will eat pet food given the opportunity, and are known to raid vegetable gardens and compost bins.
- They are more likely to be hit and killed by a car than killed by natural predators.

Virginia Opossum

Virginia Opossum

Didelphis virginiana

What is their preferred habitat?
Woodland areas along streams, and marshes and swamps.

How big are they?
About the size of a house cat, usually between 10 and 14 pounds.

What do they like to eat?
They are omnivores, and eat a wide variety of foods including fruit, grain, insects, especially ticks, earthworms, and small mammals.

What is their average lifespan in the wild?
Normally 1 to 2 years.

What is their fur color?
Gray on their bodies, with a whitish face and dark ears.

Fun Facts

- The Virginia opossum is the only marsupial found in North America, and as a species has been in existence for at least 70 million years.
- They have been known to have as many as 25 babies, but the typical litter is between 7 to 8. Their babies, once born, climb into the mother's pouch where they find teats for milk.
- During the autumn months, persimmons are one of their favorite foods. They are a highly beneficial species to humanity and consume great quantities of ticks as part of their diet. They eat up to 5,000 ticks in one week!
- They are nocturnal and sleep in hollow logs, brush piles, and other dry places during the day.
- Virginia opossums, or "possums" as they are commonly known, are famous for reacting to threats by pretending to be dead, thus discouraging attacks by predators. This reaction is involuntary and triggered by extreme fear. If overly threatened, they will enter a near coma and lie on their side, mouth and eyes open, tongue hanging out, and their breathing rate slowing by 30% or more. This death feigning normally stops when the threat is withdrawn, but can sometime last up to 4 hours.
- They are immune to rattlesnake and scorpion bites.
- They have a prehensile tail that is adapted for grasping and wrapping around tree limbs.

White-tailed Deer

White-tailed Deer

Odocoileus virginianus

What is their preferred habitat?

During summer, fields and meadows with bordering oak and hickory forests for shade. During the winter, preferably forests with coniferous stands that provide protection from the elements.

How big are they?

Males usually weigh between 100 to 300 pounds, and females between 120 and 250 pounds.

What do they like to eat?

They are herbivores, with a diet of grass, leaves, twigs, fruit, and nuts.

What is their average lifespan in the wild?

Average lifespan is between 3 and 6 years. Maximum lifespan in the wild is 20 years.

What is their fur color?

Reddish brown in the spring and summer, and gray-brown in the fall and winter. Undersides and throats are white.

Fun Facts

- Male deer are called *bucks*, females are known as *does*, and young deer as *fawns*.
- Fawns are born with white spots as camouflage but lose them within a year of birth. Fawns also take their first steps with 30 minutes of their birth. Typically, fawns stay with their mothers for about a year.
- The bucks use their antlers, which are shed and regrown yearly, as defensive weapons and in mating battles with other bucks.
- They can run up to 40 miles per hour in bursts and leap as high as 9 feet, easily clearing any fence.
- They are also excellent swimmers and often enter water to escape predators, or to reach new feeding grounds.
- When alarmed, they stomp their hooves and snort to warn other deer.
- They have a large tail that is white underneath and when threatened, the white-tailed deer raise their tail like a bright white flag.
- One out of every 3,000 will have white fur with large brown patches. Not albinos, these variations are referred to as Piebald deer. The author, in 2024, saw one of these rare creatures in the mountains of Western North Carolina.

Wild Boar

Wild Boar

Sus scrofa

What is their preferred habitat?

Dense vegetation and woodlands near accessible water.

How big are they?

Usually between 100 to 200 pounds but can grow to over 400 pounds.

What do they like to eat?

They are omnivores and prefer crops, fruit, nuts, roots, and green plants. They also will consume bird eggs, small wild mammals, young livestock, and carrion.

What is their average lifespan in the wild?

Normally 1 to 2 years.

What is their fur color?

Brown, dark gray or reddish brown, with many variations.

Fun Facts

- The wild boar is a species of wild pig native to the forests of Europe, Asia, and parts of Africa that was introduced into America.
- They have extremely poor eyesight, compensated for by an incredibly acute sense of smell and excellent hearing.
- They live in large groups, called *sounders*, made up of 6 to 20 closely related females and young. These groups can have as many as 100 members.
- They are considered an invasive species in the Northern Appalachian Mountains, as well as other areas in North America. Their habit of rooting and digging destructively in the soil, and raiding farm crops for food, are the main complaints about wild boars.
- Male wild boars can have razor sharp tusks as long as 5 inches that they use for defense, or in establishing dominance over other males.

Woodland Jumping Mouse

Woodland Jumping Mouse

Napaeozapus insignis

Meadow Jumping Mouse

Zapus hudsonius

What is their preferred habitat?

Woodland jumping mice *(shown in illustration)* prefer woodlands and forests with cool moist areas and dense vegetation. Meadow jumping mice prefer moist grasslands and avoid heavily wooded areas.

How big are they?

Woodland jumping mice are usually 7 to 9 inches long including the tail, while meadow jumping mice are usually between 8 to 10 inches long.

What do they like to eat?

Both species like to eat seeds primarily, but also east fruit, fungi, nuts, leaves, and insects.

What is their average lifespan in the wild?

Normally 1–3 years for both species.

What is their fur color?

White on the belly, dark yellowish-orange to reddish-brown on the sides, and a darker brown stripe down their back. Woodland jumping mice usually have a white-tipped tail.

Fun Facts

- Both species are similar in appearance and size but the meadow jumping mouse is duller in color and its tail is not usually tipped with white.
- Both species are recognized by their very long tails and long hind feet. They are extraordinary jumpers when frightened, jumping an astonishing 9 feet or more, like miniature kangaroos!
- Woodland jumping mice shelter in burrows they dig, and are normally active during the night, but also can be seen on cloudy or rainy days. When they are in their burrows, they typically cover the entrances with leaves or other vegetation. Their burrows are lined with grass, reeds, and leaves.
- Both species can swim both underwater and on the surface of water for short distances.

Field Notes and Observations

Field Notes and Observations

www.ingramcontent.com/pod-product-compliance
Lightning Source LLC
LaVergne TN
LVHW070856210926
848482LV00016B/947